ENERGY SECTOR STANDARD
OF THE PEOPLE'S REPUBLIC OF CHINA

中华人民共和国能源行业标准

Code for Design of Desilting Basin
for Hydropower Projects

水电工程沉沙池设计规范

NB/T 10390-2020

Replace DL/T 5107-1999

Chief Development Department: China Renewable Energy Engineering Institute
Approval Department: National Energy Administration of the People's Republic of China
Implementation Date: February 1, 2021

China Water & Power Press
中国水利水电出版社
Beijing 2024

All rights reserved. No part of this publication may be reproduced, stored in a retrieval system, or transmitted in any form or by any means—electronic, mechanical, photocopying, recording or otherwise, without prior written permission of the publisher.

图书在版编目（CIP）数据

水电工程沉沙池设计规范：NB/T 10390-2020 = Code for Design of Desilting Basin for Hydropower Projects (NB/T 10390-2020) : 英文 / 国家能源局发布. -- 北京：中国水利水电出版社, 2024. 8. -- ISBN 978-7-5226-2722-9

Ⅰ. TV673-65

中国国家版本馆CIP数据核字第2024202Q5M号

ENERGY SECTOR STANDARD
OF THE PEOPLE'S REPUBLIC OF CHINA
中华人民共和国能源行业标准

Code for Design of Desilting Basin for Hydropower Projects
水电工程沉沙池设计规范
NB/T 10390-2020
Replace DL/T 5107-1999
（英文版）

Issued by National Energy Administration of the People's Republic of China
国家能源局　发布
Translation organized by China Renewable Energy Engineering Institute
水电水利规划设计总院　组织翻译
Published by China Water & Power Press
中国水利水电出版社　出版发行
　　Tel: (+ 86 10) 68545888　68545874
　　sales@mwr.gov.cn
　　Account name: China Water & Power Press
　　Address: No.1, Yuyuantan Nanlu, Haidian District, Beijing 100038, China
　　http: //www.waterpub.com.cn
中国水利水电出版社微机排版中心　排版
北京中献拓方科技发展有限公司　印刷
184mm×260mm　16开本　3.5印张　111千字
2024年8月第1版　2024年8月第1次印刷

Price（定价）：￥580.00

Introduction

This English version is one of China's energy sector standard series in English. Its translation was organized by China Renewable Energy Engineering Institute authorized by National Energy Administration of the People's Republic of China in compliance with relevant procedures and stipulations. This English version was issued by National Energy Administration of the People's Republic of China in Announcement [2023] No. 5, dated October 11, 2023.

This version was translated from the Chinese Standard NB/T 10390-2020, *Code for Design of Desilting Basin for Hydropower Projects*, published by China Water & Power Press. The copyright is reserved by National Energy Administration of the People's Republic of China. In the event of any discrepancy in the implementation, the Chinese version shall prevail.

Many thanks go to the staff from the relevant standard development organizations and those who have provided generous assistance in the translation and review process.

For further improvement of the English version, any comments and suggestions are welcome and should be addressed to:

China Renewable Energy Engineering Institute
No. 2 Beixiaojie, Liupukang, Xicheng District, Beijing 100120, China
Website: www.creei.cn

Translating organization:

POWERCHINA Chengdu Engineering Corporation Limited

Translating staff:

ZHU Wanqiang	MENG Fanli	ZHANG Gongping	ZHANG Zhengxiang
TANG Zhidan	HU Maoyin	LI Yaqi	ZHONG Quan
WU Baijie	LIU Bin	LI Wanjun	YANG Minggang
PAN Liru	LEI Yifan		

Review panel members:

GUO Jie	POWERCHINA Beijing Engineering Corporation Limited
YAN Wenjun	Army Academy of Armored Forces, PLA
QI Wen	POWERCHINA Beijing Engineering Corporation Limited
YE Bin	POWERCHINA Huadong Engineering Corporation Limited

LI Zhongjie	POWERCHINA Northwest Engineering Corporation Limited
SHAO Guojian	Hohai University
LI Guobin	Nanjing Hydraulic Research Institute
LI Shisheng	China Renewable Energy Engineering Institute

National Energy Administration of the People's Republic of China

翻译出版说明

本译本为国家能源局委托水电水利规划设计总院按照有关程序和规定，统一组织翻译的能源行业标准英文版系列译本之一。2023年10月11日，国家能源局以2023年第5号公告予以公布。

本译本是根据中国水利水电出版社出版的《水电工程沉沙池设计规范》NB/T 10390—2020 翻译的，著作权归国家能源局所有。在使用过程中，如出现异议，以中文版为准。

本译本在翻译和审核过程中，本标准编制单位及编制组有关成员给予了积极协助。

为不断提高本译本的质量，欢迎使用者提出意见和建议，并反馈给水电水利规划设计总院。

地址：北京市西城区六铺炕北小街2号
邮编：100120
网址：www.creei.cn

本译本翻译单位：中国电建集团成都勘测设计研究院有限公司

本译本翻译人员： 朱万强　孟凡理　张公平　张正香
　　　　　　　　　唐志丹　胡茂银　李雅琦　钟　权
　　　　　　　　　吴佰杰　刘　斌　李万军　杨明刚
　　　　　　　　　潘礼儒　雷艺繁

本译本审核人员：

郭　洁　中国电建集团北京勘测设计研究院有限公司
闫文军　中国人民解放军陆军装甲兵学院
齐　文　中国电建集团北京勘测设计研究院有限公司
叶　彬　中国电建集团华东勘测设计研究院有限公司
李仲杰　中国电建集团西北勘测设计研究院有限公司
邵国建　河海大学
李国斌　南京水利科学研究院
李仕胜　水电水利规划设计总院

国家能源局

Announcement of National Energy Administration of the People's Republic of China
[2020] No. 5

National Energy Administration of the People's Republic of China has approved and issued 502 energy sector standards including *Technical Code for Real-Time Ecological Flow Monitoring Systems of Hydropower Projects* (Attachment 1) and the English version of 35 energy sector standards including *Series Parameters for Horizontal Hydraulic Hoist (Cylinder)* (Attachment 2).

Attachments: 1. Directory of Sector Standards
2. Directory of English Version of Sector Standards

National Energy Administration of the People's Republic of China

October 23, 2020

Attachment 1:

Directory of Sector Standards

Serial number	Standard No.	Title	Replaced standard No.	Adopted international standard No.	Approval date	Implementation date
...						
6	NB/T 10390-2020	Code for Design of Desilting Basin for Hydropower Projects	DL/T 5107-1999		2020-10-23	2021-02-01
...						

Foreword

According to the requirements of Document GNKJ [2015] No. 283 issued by National Energy Administration of the People's Republic of China, "Notice on Releasing the Development and Revision Plan of Energy Sector Standards in 2015", and after extensive investigation and research, summarization of practical experience, consultation of relevant advanced Chinese standards, and wide solicitation of opinions, the drafting group has prepared this code.

The main technical contents of this code include: general provisions, terms, setting conditions and design criteria for desilting basin, selection of desilting basin types, layout of desilting basin, hydraulic design and main dimension calculation, structural design, operation design, and sediment monitoring design.

The main technical contents revised are as follows:

— Adding the calculation requirements related to the structural design of desilting basins.

— Modifying the scope of application of this code, from desilting basins of Grade 3 and above of large- and medium-sized hydropower and water conservancy projects to desilting basins of hydropower projects.

— Modifying the setting criteria for desilting basins.

— Deleting the setting criteria for desilting basins of water conservancy projects.

National Energy Administration of the People's Republic of China is in charge of the administration of this code. China Renewable Energy Engineering Institute has proposed this code and is responsible for its routine management. Energy Sector Standardization Technical Committee on Hydropower Investigation and Design is responsible for the explanation of specific technical contents. Comments and suggestions in the implementation of this code should be addressed to:

China Renewable Energy Engineering Institute
No. 2 Beixiaojie, Liupukang, Xicheng District, Beijing 100120, China

Chief development organization:

POWERCHINA Chengdu Engineering Corporation Limited

Chief drafting staff:

ZHU Wanqiang   HUANG Yankun   HAO Yuanlin   JIANG Dequan

HE Xianpei	LIU Ding	TIAN Xun	LYU Jinbo
LU Peican	YUAN Liyi	CHEN Qiang	JU Lin

Review panel members:

DANG Lincai	FANG Guangda	ZHANG Hongwu	LIU Xingnian
LI Guobin	LONG Qihuang	CAO Yubo	WANG Yi
GUAN Lihai	HU Wangxing	JIANG Hongjun	WU Caiping
YANG Jinying	ZHAO Yi	LIU Rongli	LI Shisheng

Contents

1	General Provisions	1
2	Terms	2
3	Setting Conditions and Design Criteria for Desilting Basin	4
3.1	Basic Data	4
3.2	Determining Inflow Sediment Concentration and Particle Size Distribution	4
3.3	Setting Conditions and Settling Criteria for Desilting Basin	5
4	Selection of Desilting Basin Types	7
5	Layout of Desilting Basin	10
5.1	General Requirements	10
5.2	Periodic Flushing Desilting Basin	11
5.3	Continuous Flushing Desilting Basin	12
5.4	Desilting Channel	12
6	Hydraulic Design and Main Dimension Calculation	14
6.1	General Requirements	14
6.2	Approach Channel and Upstream Transition Section	14
6.3	Working Section	14
6.4	Downstream Transition Section and Water Conveyance Channel	18
6.5	Hydraulic Calculations for Desilting Basin and Sediment Flushing Channel	18
7	Structural Design	20
7.1	General Requirements	20
7.2	Structural Calculation	21
8	Operation Design	22
9	Sediment Monitoring Design	23
Appendix A	Sediment Settling Calculation	25
Appendix B	Calculation on Suspended Sediment Settling Velocity	32
Appendix C	Hydraulic Calculation on Sediment Releasing for Desilting Basin	34
Appendix D	Hydraulic Calculation on Sediment Flushing for Desilting Basin	38
Explanation of Wording in This Code		43
List of Quoted Standards		44

1 General Provisions

1.0.1 This code is formulated with a view to standardizing the design of desilting basins for headrace structures of hydropower projects.

1.0.2 This code is applicable to the design of hydropower project desilting basin for removing suspended load.

1.0.3 In addition to this code, the design of desilting basin for hydropower projects shall comply with other current relevant standards of China.

2 Terms

2.0.1 desilting basin

structure for settling suspended load with particles larger than the design sedimentation size in sediment-laden flow to reduce sediment concentration

2.0.2 periodic flushing desilting basin

desilting basin in which the deposited sediments are alternately settled and flushed out through flushing sluice

2.0.3 continuous flushing desilting basin

desilting basin in which deposited sediments are flushed into the downstream river course under continuous water flow

2.0.4 desilting channel

long, wide and shallow soil channel formed by using natural depressions

2.0.5 working section

main desilting basin section for sediment settling

2.0.6 working length

length of the working section

2.0.7 working width

width of the working section

2.0.8 working depth

water depth from the normal water level to the design deposition elevation in a desilting basin

2.0.9 working flow

available discharge for the hydropower stations with periodic flushing desilting basin; sum of the available discharge and the sediment flushing discharge from the bottom outlet at the basin end for the periodic flushing desilting basin with sediment flushing outlet; sum of the available discharge and the flushing discharge for the continuous flushing desilting basin

2.0.10 flushing flow

discharge from the inlet sluice into the basin chamber (tank) of periodic flushing desilting basin for sediment flushing

2.0.11 basin chamber

troughs formed by partitioning working section in flow direction using walls higher than the normal water level, according to the working flow, working width and working depth

2.0.12 basin tank

compartments formed by partitioning periodic flushing desilting basin chamber in flow direction using walls higher than the flushing level above the design deposition surface, according to the flushing flow

2.0.13 overflow area

overflow weir area set at the end of the working section in a periodic flushing desilting basin to withdraw surface water

2.0.14 mean annual sediment concentration through turbine

ratio of multi-year sediment runoff to multi-year water flow through the turbine

2.0.15 coarse sediment concentration

suspended sediment concentration with particles larger than the design minimum sedimentation size

3 Setting Conditions and Design Criteria for Desilting Basin

3.1 Basic Data

3.1.1 The desilting basin design shall collect the basic data as follows:

1. Topography and geology.

2. Measured flow, suspended sediment concentration, particle size distribution and its analysis method, and water temperature in the reach where the project is located.

3. Mineral composition and hardness of suspended load.

4. The rating curve in the natural river at the outlet of the flushing channel of the desilting basin.

5. The water surface profile of each level of river flow, where the basin sidewall is close to the natural river.

6. Ice regime and air temperature.

7. Sources and properties of pollutants in the river.

3.1.2 The desilting basin design shall collect the following design data of relevant specialties:

1. Water levels of the approach channel and conveyance channel to the desilting basin.

2. Design runoff of the river reach where the project is located.

3. Design available discharge.

4. Sediment-laden capacity of flow in approach channel and water conveyance channel.

5. Abrasion resistance of components, measures against abrasion, overhaul interval and operation head of turbine.

6. Operating modes for reservoir and hydropower station.

3.2 Determining Inflow Sediment Concentration and Particle Size Distribution

3.2.1 The time-framed average inflow sediment concentration shall be calculated according to the daily average sediment concentration and the daily average flow of the intake structure by the following formula:

$$S = \frac{\sum_{i=1}^{T} Q_i S_i}{\sum_{i=1}^{T} Q_i} \qquad (3.2.1)$$

where

S is the time-framed average inflow sediment concentration (kg/m³);

T is the number of days in the time frame;

Q_i is the average inflow on the ith day (m³/s);

S_i is the daily average sediment concentration on the ith day (kg/m³).

3.2.2 The representative inflow suspended sediment concentration and particle size distribution as the design basis shall be determined through analysis and demonstration according to the desilting basin type, sediment transport characteristics and reservoir deposition behavior.

3.3 Setting Conditions and Settling Criteria for Desilting Basin

3.3.1 The preliminary judgement on the setting of desilting basin (Figure 3.3.1) shall be determined according to the intersection of the mean annual sediment concentration through turbine S_p or the mean annual coarse sediment concentration through turbine S'_p and the rated head of turbine H_r. If the intersection falls in the area A of the figure, desilting basin need not be provided; if the intersection falls in the area C, desilting basin should be provided; and if the intersection falls in the area B, desilting basin need not be provided if reliable measures against abrasion are taken in the hydraulic design, structural design and materials for the components of hydraulic turbine.

3.3.2 On the basis of preliminary judgement, the setting of desilting basin shall be determined through techno-economic comparison according to the analysis results of abrasion and damage to turbine due to sediment, considering the role of the hydropower station in the system, project layout, investment for desilting basin, sediment characteristics, abrasion resistance of turbine, hydropower station benefits and operation & maintenance, etc.

3.3.3 In desilting basin setting, the settling rate should be taken as 80 % to 85 % for the sediment with particles larger than and equal to the design minimum sedimentation size, and the design minimum sedimentation size for the desilting basin may be determined by the rated head of turbine as per Table 3.3.3.

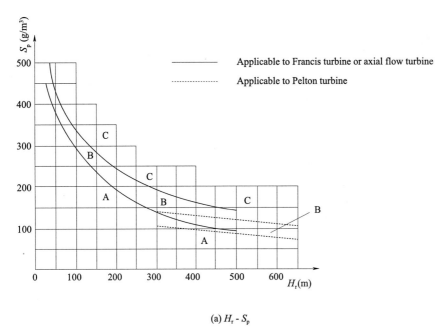

(a) H_r - S_p

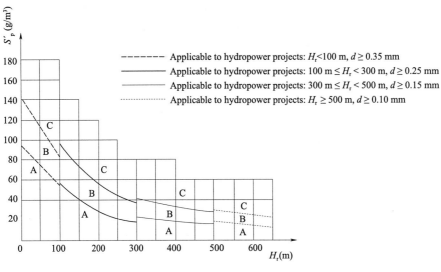

(b) H_r - S'_p

Figure 3.3.1 Preliminary judgement on setting of desilting basin

Table 3.3.3 Design minimum sedimentation size for desilting basin

Rated head H_r (m)	$H_r < 100$	$100 \leq H_r < 300$	$300 \leq H_r < 500$	$H_r \geq 500$
Design minimum sedimentation size (mm)	0.35	0.25	0.15	0.1

4 Selection of Desilting Basin Types

4.0.1 For hydropower projects, hydraulic flushing or non-flushing desilting basin may be selected according to topographical and water head conditions. Periodic flushing desilting basin (Figure 4.0.1-1) or continuous flushing desilting basin (Figure 4.0.1-2) may be selected for hydraulic flushing desilting basins, and continuous flushing desilting basins may adopt single-chamber basin or multi-chamber basin. Deposition or mechanical dredging desilting basin may be selected for non-flushing desilting basins. Other types of desilting basins may also be selected as required by the projects.

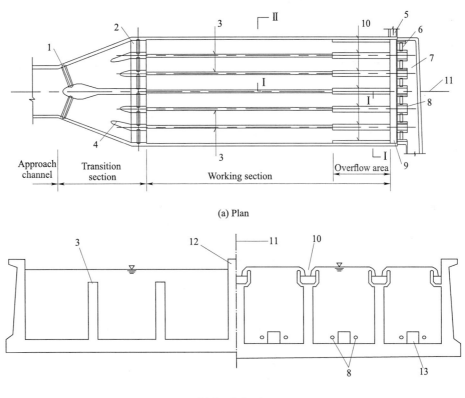

Key

1 basin chamber inlet sluice
2 basin tank inlet sluice
3 partition in basin tank
4 baffle wall
5 water conveyance channel
6 flushing sluice
7 sediment flushing channel
8 sediment flushing outlet
9 transverse trough
10 lateral trough
11 desilting basin centerline
12 basin chamber partition
13 flushing sluice outlet

Figure 4.0.1-1　Periodic flushing desilting basin

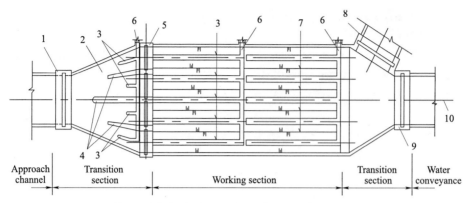

(a) Plan of single-chamber continuous flushing desilting basin

Key

1 inlet sluice
2 expansion section
3 sediment flushing branch gallery
4 baffle wall
5 tranquiller racks section
6 main flushing gallery
7 trapezoidal trough wall
8 emergency flushing sluice
9 outlet sluice
10 desilting basin centerline

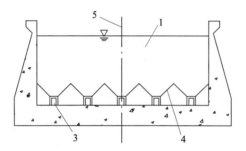

(b) Single-chamber cross section

Key

1 settling chamber
3 branch flushing gallery
4 trapezoidal trough wall
5 desilting basin centerline

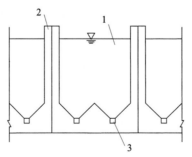

(c) Multi-chamber cross section

Key

1 settling chamber
2 partition
3 branch flushing gallery

Figure 4.0.1-2 Continuous flushing desilting basin

4.0.2 Hydraulic flushing desilting basins shall have sufficient water head and flow for sediment flushing. Periodic flushing desilting basin should be applied to wide-open terrain, while continuous flushing desilting basin to narrow terrain.

4.0.3 Mechanical dredging type desilting basin should be adopted for the hydropower projects with insufficient sediment flushing water head, and there shall be sufficient sediment dumping sites which shall not adversely affect surroundings and the project.

4.0.4 Desilting channel (Figure 4.0.4) should be selected for low-lying areas in plains. The pre-excavated desilting channel, pumping-inflow desilting channel or the combination of the pumping-inflow and gravity flowing constructed in stages may be selected according to the topographical and operating conditions.

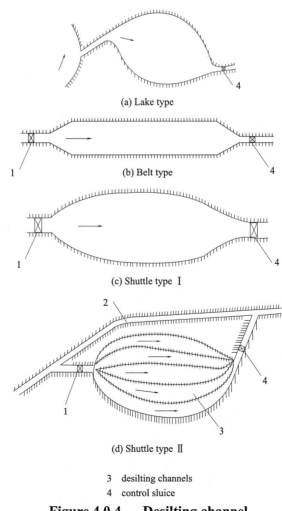

Key
1 inlet sluice
2 channel
3 desilting channels
4 control sluice

Figure 4.0.4   Desilting channel

5 Layout of Desilting Basin

5.1 General Requirements

5.1.1 The layout of desilting basin shall be coordinated with the general project arrangement, utilizing the terrain reasonably and avoiding unfavorable areas, considering geological conditions, engineering characteristics, operation conditions, economy, etc.

5.1.2 The inlet of desilting basin should be arranged in the stable reach with smooth flow. The river channel may be appropriately improved if necessary.

5.1.3 The desilting basin should be arranged close to the intake of the headworks, if the site condition or flushing head fails the requirements, it may be arranged downstream the headrace to an appropriate location; it may also be arranged underground.

5.1.4 The axis of the desilting basin should coincide with that of the upstream approach channel, and when there is an included angle between these two axes, engineering measures shall be taken to ensure an even distribution of the transverse and vertical velocity of the water into the working section of the desilting basin.

5.1.5 The setting of the main components of desilting basin shall be such as to make the flow evenly diffuse to the working section of the desilting basin, and shall meet the following requirements:

1. The approach channel of desilting basin shall meet the requirements of water conveyance and deposition control under the design discharge. Bed load barrier or flushing facilities should be provided at an appropriate position of the approach channel.

2. The expansion section should adopt symmetrical layout, and the flaring angle should not be greater than 12° at one side. If an asymmetric layout is adopted, the sum of the flaring angles at both sides should not be greater than 24°. Sudden drop should be avoided at the connection between the expansion section and working section.

3. Baffle wall shall be arranged in the expansion section, and the position, size, direction, row spacing and interval of tranquiller racks should be determined by hydraulic model tests.

4. The sill top level of the inlet sluice for the basin chamber and basin tank shall be flush with or slightly higher than the upstream bed, and should be higher than the design sediment siltation surface of the basin

chamber and basin tank. If there are trashes, the inlet sluice shall be equipped with the trash rack and cleaning facilities. The gate shall meet the requirements of partial opening operation.

5   The side wall and partition of the basin chamber shall be smoothly connected with the pier of the inlet sluice of the corresponding basin chamber, and the slope of water-facing surface of the side wall and partition should not be less than 1 : 0.1.

6   Lateral overflow weirs may be provided at appropriate locations of approach channel, working section or water conveyance channel to the desilting basin if necessary, and the weir crest level should be slightly higher than the operating water level of the desilting basin. The discharge capacity of lateral overflow weirs shall be determined according to the operating requirements, etc.

5.1.6 Non-pressure flow should be adopted for sediment flushing channel, and the longitudinal slope shall not be less than that of the desilting basin working section. Abrasion resistant and siltation control measures shall be taken at the outlet of the sediment flushing channel to ensure a smooth sediment flushing at the flood level of a 2-year return period.

5.2   Periodic Flushing Desilting Basin

5.2.1 Basin chamber inlet sluices shall be provided for the periodic flushing desilting basin. Basin tank inlet sluices shall be provided when a basin chamber is divided into basin tanks, and the number of sluice bays shall be consistent with that of basin tanks. The partition between basin tanks shall be connected with the inlet sluice pier of the corresponding basin tank, and the top of the partition shall be 0.5 m higher than the elevation of the sediment flushing water surface above the design sediment siltation surface.

5.2.2 Lateral and transverse overflow weirs and water collecting channel may be provided at the working section end of the desilting basin. The flow depth at the overflow weir crest should be less than 0.2 m, and both the overflow weir and the water collecting channel shall allow for free outflow. The profile of lateral and transverse water collecting channel shall be determined by hydraulic calculation.

5.2.3 Sediment flushing outlets should be provided at the bottom of the end wall of the desilting basin, and the flushing discharge of the outlets should be 5 % to 8 % of the working flow. The number of outlets in each basin tank should not be less than 2.

5.2.4 Sediment flushing sluices shall be installed at the working section end of the desilting basin, and the gate and hoist of sediment flushing sluice shall allow for partial opening operation.

5.2.5 The outlet gates shall be provided for the downstream water conveyance channel of the desilting basin.

5.3 Continuous Flushing Desilting Basin

5.3.1 The flushing system of the desilting basin shall have several branch galleries and main galleries.

5.3.2 The bottom of the working section of the desilting basin may be constructed into a number of inverted trapezoidal troughs along the desilting basin width, and the included angle between the trough wall and the horizontal direction should be greater than the underwater repose angle of sediment.

5.3.3 The branch galleries shall be arranged below the trapezoidal troughs of the desilting basin along the flow, and the bottom of the trapezoidal troughs shall have sediment feed holes connected with the branch galleries. Each branch gallery should connect to a main gallery. Multiple flushing systems should be provided in the working section of the desilting basin, and the length of the branch gallery of the upstream flushing system should be shorter than that of the downstream flushing system.

5.3.4 The main and branch galleries shall meet the hydraulic requirements for sediment flushing, and the abrasion-resistant measures shall be taken. Abrasion resistant and siltation control measures shall be taken at the outlets of the main flushing galleries to ensure a smooth sediment flushing.

5.3.5 For single-chamber continuous flushing desilting basins, emergency flushing sluice should be provided at the end of the working section. The gate and hoist of the sediment flushing sluice shall allow for partial opening operation.

5.3.6 Gradually contracting structure should be adopted for the downstream transition section of the desilting basin.

5.4 Desilting Channel

5.4.1 The desilting channel may be of lake, belt or shuttle type depending on topographical conditions.

5.4.2 Control gates should be set at the inlet and outlet of the desilting channel and be arranged in upstream, midstream and downstream sections respectively.

5.4.3 The slope of the embankment and partition dikes of the desilting channel shall be determined according to the properties of the embankment construction materials and the scale of the desilting channels.

5.4.4 The crest elevation of the embankment shall be determined by the water surface elevation plus wave run-up and freeboard.

5.4.5 The crest width of the embankment shall meet the requirements for traffic and protection.

5.4.6 Effective seepage prevention, cut-off and drainage measures shall be taken in the layout of desilting channel.

6 Hydraulic Design and Main Dimension Calculation

6.1 General Requirements

6.1.1 The dimension alternatives of the main hydraulic sections of a desilting basin shall be proposed according to topographical and geological conditions, and be finally selected through techno-economic comparison.

6.1.2 The periodic flushing desilting basin shall mainly consist of upstream transition section, working section and its overflow area, water collection and sediment flushing facilities.

6.1.3 The continuous flushing desilting basin shall mainly consist of upstream transition section, working section, downstream transition section and sediment flushing galleries.

6.1.4 The hydraulic calculations for desilting basin should include the following:

1 Flow capacity calculation.

2 Water surface profile calculation.

3 Sediment settling rate calculation.

4 Sediment deposition calculation.

5 Sediment flushing calculation.

6.1.5 The sediment settling calculation shall comply with Appendix A of this code.

6.1.6 The desilting basins with complex hydraulic conditions should be verified by model tests.

6.2 Approach Channel and Upstream Transition Section

6.2.1 The hydraulic design of approach channel and upstream transition section shall be such that the flow is smooth and stable in the channel, the water surface fluctuation decreases gradually, and backflow and vortex are avoided.

6.2.2 The design flow velocity in the transition section of the approach channel shall be greater than the flow velocity without silting of suspended load and less than the flow velocity without abrasion of the channel.

6.3 Working Section

6.3.1 The dimensions of working section of a desilting basin shall be comprehensively determined by the hydraulic factors, e.g., inflow, sediment settling velocity, volume and time of sediment deposition, critical sediment-flushing flow velocity, flushing cycle, sediment flushing mode, etc.

6.3.2 For periodic flushing desilting basins, the working section inlet shall have adequate working depth and take into account the influence of sediment deposition during operation. The flow depth at the working section inlet should not be excessive. The working depth and the flow depth at the working section inlet can be calculated by the following formulae:

$$H_e = H - \Delta H_k \tag{6.3.2-1}$$

$$H \leq \Delta Z + \frac{q}{v_c} - (iL_w + i_0 L_0) \tag{6.3.2-2}$$

where

H_e is the working depth at the working section inlet (m);

H is the flow depth at the working section inlet (m);

ΔH_k is the design sediment deposition thickness at the working section inlet (m), which may be taken as 25 % H to 30 % H in the preliminary scheme;

ΔZ is the difference between the water level at the working section inlet and the natural river water level at the outlet of the sediment flushing channel (m);

q is the sediment flushing flow per unit width of the sediment flushing channel [m³/(s · m)];

v_c is the sediment-flushing flow velocity at the outlet of the sediment flushing channel (m/s);

i is the bottom slope of the working section of the desilting basin;

L_w is the length of the working section of the desilting basin (m);

i_0 is the bottom slope of the sediment flushing channel;

L_0 is the length of the sediment flushing channel (m).

6.3.3 The working depth of working section inlet of the continuous flushing desilting basin shall ensure sufficient flow velocity for sediment flushing, which can be calculated by the following formula:

$$H_e \leq \Delta Z_1 - (1+\Sigma\xi)\frac{v_c^2}{2g} - v_c^2 \int_0^L \frac{dl}{C^2 R} \tag{6.3.3}$$

where

ΔZ_1 is the difference between the operating water level at the desilting basin and the top level of the gallery outlet (m);

$\Sigma \xi$ is the sum of local head loss coefficients;

L is the total length of branch galleries and main galleries (m);

g is the acceleration of gravity (m/s²);

C is the Chezy coefficient;

R is the hydraulic radius (m).

6.3.4 The flow depth in the desilting channel should be 2.0 m to 3.5 m.

6.3.5 The width of single-chamber working section of hydraulic flushing desilting basins is determined by flow, flow depth and flow velocity, and can be calculated by the following formula:

$$B = \frac{Q}{H_w \upsilon} \quad (6.3.5)$$

where

B is the working width (m);

Q is the working flow (m³/s);

υ is the average flow velocity of working section (m/s); in preliminary scheme, it may be taken as 0.05 m/s to 0.15 m/s when the minimum particle size of deposited sediment is 0.05 mm to 0.10 mm, 0.25 m/s to 0.55 m/s when the minimum particle size is 0.25 mm, and 0.40 m/s to 0.80 m/s when the minimum particle size is 0.35 mm;

H_w is the average working depth (m).

6.3.6 The ratio of width to depth of the single-chamber working section in a hydraulic flushing desilting basin should not be greater than 4.5.

6.3.7 The width of the tank in the working section of a periodic flushing desilting basin can be calculated by the following formula:

$$b_s = \frac{Q'_s}{q_s} \quad (6.3.7)$$

where

b_s is the width of a tank in the working section of a periodic flushing desilting basin (m);

Q'_s is the flushing flow of the tank (m³/s);

q_s is the flushing flow per unit width of the tank [m³/(s · m)].

6.3.8 The working width of the desilting channel may be preliminarily selected by Formula (6.3.5) of this code. When the allowable particle size of the outflow sediment from the basin is not greater than 0.05 mm, v may be taken as 0.20 m/s to 0.40 m/s.

6.3.9 The determination of the working length of a desilting basin shall meet the following requirements:

1. Alternative schemes shall be proposed for main dimensions, and the working length shall be determined by sediment settling calculation in accordance with Appendix A of this code.

2. The calculation on suspended sediment settling velocity shall comply with Appendix B of this code.

3. The design length of the working section of a desilting basin should be 1.2 times the calculated length.

6.3.10 For the periodic flushing desilting basin with lateral overflow weir, the grouped sediment concentration of outflow may be calculated based on the vertical distribution of sediment concentration in front of the overflow weir at the working section end. The total length of the lateral overflow weir of basin chambers can be calculated by the following formula:

$$L_e = \frac{1}{\beta}\left(\frac{Q_f}{m\sqrt{2g}h_f^{3/2}} - b_f\right) \tag{6.3.10}$$

where

L_e is the total length of the lateral overflow weir of the basin chamber (m);

Q_f is the available flow of the basin chamber (m³/s);

b_f is the width of the transverse overflow weir of the basin chamber (m);

h_f is the average water depth at the weir crest (m);

m is the flow coefficient;

β is the lateral influence coefficient of the overflow weir.

6.3.11 The grouped sediment concentration of the desilting channel and outflow section at the end of each time frame may be calculated in accordance with Section A.2 of this code. When the particle size distribution and sediment concentration of the outflow section are close to the upper limit of the design criteria, the cumulative operating time may be taken as the service life of the desilting channel.

6.3.12 The working section of a periodic flushing desilting basin shall have such a longitudinal bottom slope that satisfies the requirements of sediment-flushing flow velocity, and the longitudinal bottom slope can be calculated by the following formula:

$$i \geq \frac{v_{1c}^2}{C^2 R} \tag{6+.3.12}$$

where

v_{1c} is the sediment-flushing flow velocity in working section (m/s).

6.4 Downstream Transition Section and Water Conveyance Channel

6.4.1 The flow depth in the transition section and water conveyance channel of pressure headrace shall meet the requirement of minimum submergence at the pressure inlet, to avoid vertical vortex and aeration.

6.4.2 The maximum water level of the transition section and the water conveyance channel shall be determined based on the maximum surge water level in the case of sudden full load rejection of the power station during normal operation at design flow.

6.4.3 Energy dissipators shall be provided for the overflow weir of desilting basins.

6.5 Hydraulic Calculations for Desilting Basin and Sediment Flushing Channel

6.5.1 The hydraulic calculation on sediment flushing for desilting basin shall comply with Appendix C of this code. The flow velocity for sediment flushing shall be greater than the critical sediment-flushing flow velocity. The critical sediment-flushing flow velocity can be calculated by the relevant formulae in Section C.1 of this code.

6.5.2 The flushing cycle and design volume for sediment deposition of periodic flushing desilting basins shall be determined through techno-economic comparison considering operations. The deposition volume, deposition duration and flushing cycle may be calculated in accordance with Appendix A of this code.

6.5.3 The design flow per unit width and duration for sediment flushing of periodic flushing desilting basins may be calculated by the hydraulic method for non-pressure sediment flushing as specified in Section C.2 of this code.

6.5.4 The flushing duration of periodic flushing desilting basins may be divided into three stages: evacuating flushing, backward flushing and flushing along the course. The flushing calculation may be carried out in accordance with Appendix D of this code.

6.5.5 The hydraulic calculation for the gallery of continuous flushing desilting basins may be carried out by the hydraulic method for pressure sediment flushing as specified in Section C.1 of this code. In the check of sediment transport capacity of a flushing gallery, the measured maximum sediment concentration may be taken as the inflow sediment concentration.

7 Structural Design

7.1 General Requirements

7.1.1 The structural type and size of a desilting basin shall be determined through techno-economic comparison according to the project layout, hydraulic design, foundation and operation conditions.

7.1.2 The desilting basin should be of reinforced concrete structure, and may also be of other structures.

7.1.3 The structure of a desilting basin shall satisfy the requirements for durability, stability and strength under various design situations. The concrete strength of the basin structure shall not be inferior to C20, the strength of the masonry structure shall not be inferior to MU30, and the strength of cement mortar shall not be inferior to M10.

7.1.4 The concrete structure shall be in accordance with the current sector standard DL/T 5057, *Design Specification for Hydraulic Concrete Structures*.

7.1.5 The cross-sectional type and size of the working section, sediment flushing gallery and sediment releasing gallery shall facilitate sediment flushing and maintenance. The wetted section of the branch and main sediment flushing galleries should be rectangular. The height of the main gallery should not be less than 1.5 m, and the wetted section of the branch galleries shall be enlarged along the course.

7.1.6 The branch gallery of a continuous flushing desilting basin should be covered with a prefabricated plate. The cover plate shall be provided with sediment feed holes which should be of louver type.

7.1.7 Measures shall be taken to prevent freezing and release ice jam for the desilting basins in cold or severe cold regions according to the operation requirements in winter and spring. Measures against frost heave shall be taken for the basin foundation.

7.1.8 The related structures of a desilting basin shall meet the requirements for safe and stable operation, and necessary treatment measures shall be taken for the foundation.

7.1.9 The desilting basin shall be equipped with necessary facilities such as the ladder, access bridge, service bridge, manhole, lifting hook, etc. according to the needs of operation, traffic and maintenance.

7.1.10 The side walls of the desilting basin shall have sufficient freeboard. The side walls of the basin chamber should be 0.5 m higher than the maximum water level, and the partition should be 0.5 m higher than the normal operating level. Drainage channels should be arranged on the ground surface outside the side walls.

7.1.11 Considering the local climate conditions, structural characteristics, foundation restraint, etc., the structure of the desilting basin shall meet the requirements for temperature deformation and settlement, and the structural joints including expansion joints and settlement joints shall be set at a spacing of 10 m to 20 m along the direction perpendicular to the flow. When a partition or baffle wall is provided in the working section or the upstream transition section, the partition or baffle wall may form a T-shaped structure with the base slab, which may be calculated as a T-shaped structure. When there is no partition in the working section, if the basin is wide, longitudinal structural joints shall be set according to the foundation conditions; if the basin is narrow, the side wall and base slab may be calculated as a U-shaped integral structure. Waterstops shall be set for structural joints.

7.2 Structural Calculation

7.2.1 The structural calculation of a desilting basin shall consider the persistent, transient and accidental design situations.

7.2.2 The structural design of a desilting basin should be subject to the following calculations and check as required:

1 The overall stability against sliding and floating of structure.

2 The foundation bearing capacity and foundation settlement.

3 The strength of each part of the structure.

7.2.3 The structure and components of the desilting basin shall be analyzed and calculated according to the natures of various loads, and the analysis and calculation may be conducted in accordance with the current sector standard NB/T 35023, *Design Code for Sluice*.

7.2.4 For the separate structures on soil foundation, tensile stress shall not occur on the foundation surface.

7.2.5 For the separate structures on rock foundation, tensile stress shall not occur on the foundation surface under persistent situation. Under transient and accidental design situations, the tensile stress not greater than 0.1 MPa is allowed on the foundation surface.

8 Operation Design

8.0.1 The operation design of the desilting basin shall consider the operating water level, design inflow, design sediment concentration, design standard of sediment settlement, operation time, etc.

8.0.2 The desilting basin shall be equipped with necessary facilities for operation, maintenance, dredging or auxiliary flushing according to the operation requirements.

8.0.3 The operation design shall propose the suggestions on maintenance time window according to the functional requirements and operation characteristics of the desilting basin.

8.0.4 For a periodic flushing desilting basin, the operation design shall propose the flushing time window, flushing mode and gate operating mode. When the inflow sediment concentration or particle size does not satisfy the design control standards, the requirement to reduce the inflow or even suspend the operation shall be proposed. The operating mode of the sediment releasing outlet shall be proposed in the operation design.

8.0.5 For a continuous flushing desilting basin, the operation design shall propose the sediment releasing mode of continuously opening the gallery outlet gate. When the inflow sediment concentration is greater than the design value, the requirement to adjust the operation arrangement or even suspend the operation shall be proposed.

8.0.6 The desilting channel should adopt the staged sediment silting of high deposition for high water level and low deposition for low water level. For long desilting channels, segmented settling from upstream to downstream may be adopted.

8.0.7 For mechanical dredging desilting basins, sediment dumping requirements shall be proposed.

9 Sediment Monitoring Design

9.0.1 The sediment monitoring design shall define the sediment observation items, observation layout, observation facilities and instruments, observation methods, observation data compilation and analysis requirements according to the project type, layout and sediment conditions.

9.0.2 The main flow and sediment observation items may include the following:

1. Inflow and water level.

2. Changes in sediment concentration and particle size distribution at the inlet, outlet and along the desilting basin.

3. Flow velocity distribution.

9.0.3 In addition to those in Article 9.0.2 of this code, the following items may be included in the observation of periodic flushing desilting basins:

1. Deposition changes and dry density of sediment.

2. Sediment flushing discharge and sediment concentration at bottom outlet.

3. Vertical distribution of sediment concentration in front of the overflow weir at the desilting basin end.

4. Changes in sediment concentration at the inlet and outlet cross sections of the basin tank, and water surface and sediment surface along the tank during flushing.

9.0.4 In addition to Article 9.0.2 of this code, the observation items for continuous flushing desilting basin may also include taking samples at the outlet of each flushing gallery to calculate sediment concentration and sediment transport rate and analyze the particle size distribution.

9.0.5 The layout and quantity of observation sections shall be able to reflect the change of hydraulic sediment factors along the desilting basin, and the facilities shall be arranged convenient for observation.

9.0.6 The observation methods for flow and water level shall comply with the current sector standard SL616, *Specification for Hydraulic Prototype Observation*, and the test method for suspended load shall comply with the current national standard GB/T 50159, *Code for Measurement of Suspended Load in Open Channels*.

9.0.7 When the observation of settling effect is proposed, the observation of sediment concentration and particle size gradation through turbine should be arranged.

9.0.8 For desilting channel, observation of flow and sediment concentration shall be carried out over the period during which the inflow and sediment concentration are relatively stable. The flow and sediment change along the course shall be considered when determining the observation time of the upstream and downstream cross sections.

Appendix A Sediment Settling Calculation

A.1 Water Surface Profile Calculation

When the working width, deposition thickness or working depth of the desilting basin change along the course, the desilting basin shall be divided into n calculation sections, and the water surface profile shall be calculated from downstream to upstream by the following formulae:

$$L_k = \frac{\left(H_{k+1} + \frac{v_{k+1}^2}{2g}\right) - \left(H_k + \frac{v_k^2}{2g}\right)}{i_k - \overline{j}_k} \qquad (A.1\text{-}1)$$

$$\overline{v}_k = \frac{v_k + v_{k+1}}{2} \qquad (A.1\text{-}2)$$

$$\overline{j}_k = \frac{\overline{v}_k^2}{\overline{C}_k^2 \overline{R}_k} \qquad (A.1\text{-}3)$$

$$\overline{R}_k = \frac{R_k + R_{k+1}}{2} \qquad (A.1\text{-}4)$$

$$\overline{C}_k = \frac{1}{n_k} \overline{R}_k^{1/6} \qquad (A.1\text{-}5)$$

where

L_k is the length of basin section k (m);

H_k, H_{k+1} are the water depths of the upstream and downstream cross sections of basin section k (m);

v_k, v_{k+1} are the flow velocities at the upstream and downstream cross sections of basin section k (m/s);

i_k is the longitudinal bed slope of basin section k;

$\overline{j}_k$ is the average hydraulic gradient of basin section k;

g is the acceleration of gravity (m/s^2);

$\overline{v}_k$ is the average flow velocity in basin section k (m/s);

R_k, R_{k+1} are the hydraulic radius of the upstream and downstream cross sections of basin section k (m);

$\overline{R}_k$ is the average hydraulic radius of basin section k (m);

$\overline{C}_k$ is the average Chezy coefficient of basin section k (m);

n_k is the roughness coefficient of basin section k;

    k is the numbering of the basin section, from the inlet of working section to the end of basin or the head of overflow area, $k = 1, 2,…,n$.

A.2 Calculation of Sediment Settling Rate

A.2.1 The calculation of sediment settling rate shall consist of the calculation of sediment concentration of each particle size group of each cross section from upstream to downstream in the basin, and then the settling rate of each particle size group of each basin section, each particle size group of whole basin and the sediment settling rate of particles larger than a certain size. The sediment concentration and settling rates may be calculated in accordance with the following requirements:

1 The sediment concentration of each particle size group in the lower cross section of the basin section can be calculated by the following formula:

$$S_{i(k+1)} = S_{ik} e^{-\alpha_{ik} \frac{\overline{\omega}_i L_k}{q_k}} \tag{A.2.1-1}$$

where

    $S_{ik}, S_{i(k+1)}$ are the grouped sediment concentrations in the upstream and downstream cross sections of basin section k (kg/m³);

    α_{ik} is the saturation recovery coefficient of particle size group i of basin section k;

    $\overline{\omega}_i$ is the average settling velocity of particle size group i (m/s);

    q_k is the discharge per unit width of basin section k [m³/(s · m)];

    i is the numbering of particle size group, ranking from small to large, $i = 1,2,…$.

2 The sediment settling rate of each particle size group of the basin section can be calculated by the following formula:

$$\eta_{ik} = \frac{S_{ik} - S_{i(k+1)}}{S_{ik}} \tag{A.2.1-2}$$

where

    η_{ik} is the sediment settling rate of particle size group i of basin section k.

3 The sediment settling rate of each particle size group of the whole basin can be calculated in accordance with the following requirements:

1) When there is no overflow area in the working section, the sediment settling rate can be calculated by the following formula:

$$\eta_i = 1 - \prod_{k=1}^{n}(1 - \eta_{ik}) \qquad (A.2.1\text{-}3)$$

2) When the overflow area is set at the end of the working section, the sediment settling rate can be calculated by the following formula:

$$\eta_i = 1 - (1 - \eta_{if})\prod_{k=1}^{n}(1 - \eta_{ik}) \qquad (A.2.1\text{-}4)$$

where

η_i is sediment settling rate of particle size group i of the whole basin;

η_{if} is sediment settling rate of particle size group i of the overflow area.

4 The sediment settling rate for the particle size larger than group i can be calculated by the following formula:

$$\eta_i = \frac{\sum_{i=m}^{l}\eta_i \Delta P_i}{\sum_{i=m}^{l}\Delta P_i} \qquad (A.2.1\text{-}5)$$

where

η_i is the sediment settling rate for the particle size larger than group i. When $l = 1$, η_1 is the total settling rate η;

l is the numbering of the lower limit value of particle size group i, ranking from large to small, $l = m, m - 1, \ldots, 2, 1$;

ΔP_i is the weight percentage of particle size group i of inflow suspended load (%).

A.2.2 The weight percentage of particle size group i of deposited sediment in the basin can be calculated by the following formula:

$$\Delta P_{di} = \frac{\eta_i \Delta P_i}{\eta} \qquad (A.2.2)$$

where

ΔP_{di} is the weight percentage of particle size group i of sediment in the basin, flushing gallery or sediment flushing outlet (%);

η is the total sediment settling rate.

A.3 Saturation Recovery Coefficient

The saturation recovery coefficient should be determined by the measured data of the existing desilting basin with similar water quality and sediment characteristics. If unavailable, it may be determined by the following methods:

1. The saturation recovery coefficient for coarse particles of suspended load can be calculated by the following formulae:

$$\alpha_{ik} = K_z \left(\frac{\bar{\omega}_i}{u_{*k}} \right)^{0.25} \tag{A.3-1}$$

$$u_{*k} = \frac{n_k v_k \sqrt{g}}{R_k^{1/6}} \tag{A.3-2}$$

where

- α_{ik} is the saturation recovery coefficient of particle size group i of basin section k;
- K_z is the comprehensive empirical coefficient. When the ratio of width to depth of the basin is 1.5 to 4.0, K_z is taken as 1.2 to 1.0;
- u_{*k} is the flow friction velocity of basin section k (m/s).

2. The sectional saturation recovery coefficient for fine particles of suspended load can be calculated by the following formula:

$$\alpha_{ik} = 6.644 \times 10^{-5} \bar{J}_k^{-0.61} \left(\frac{\bar{\omega}_i}{u_{*k}} \right)^{-0.62} \tag{A.3-3}$$

A.4 Calculation of Deposition and Flushing Cycle

A.4.1 For periodic flushing desilting basins:

1. The daily deposition volume can be calculated by the following formula:

$$V_d = \frac{86.4 \bar{S}_0 Q \eta}{\rho_d} \tag{A.4.1-1}$$

where

- V_d is the daily deposition volume (m³/d);
- $\bar{S}_0$ is the average sediment concentration during the design deposition period (kg/m³);
- Q is the working flow (m³/s);
- ρ_d is the dry density of deposition (t/m³).

2 The flushing cycle can be calculated by the following formula:

$$T_0 = \frac{V_0}{V_d} \tag{A.4.1-2}$$

where

T_0 is the design flushing cycle (d);

V_0 is design deposition volume (m³).

A.4.2 For periodic flushing desilting basin, the deposition volume and flushing cycle may also be calculated as follows:

1 The deposition volume of the working section k without overflow area in a certain time frame can be calculated by the following formulae:

$$V_k = \frac{86.4 Q t_j (S_k - S_{k+1})}{\rho_d} \tag{A.4.2-1}$$

$$S_k = \sum_{i=1}^{m} S_{ik} \tag{A.4.2-2}$$

$$S_{k+1} = \sum_{i=1}^{m} S_{i(k+1)} \tag{A.4.2-3}$$

where

V_k is the deposition volume of basin section k (m³);

t_j is the calculation duration (d);

S_k, S_{k+1} are sediment concentrations in upstream and downstream cross sections of basin section k (kg/m³).

2 The average deposition thickness of basin section k can, based on the deposition volume, be calculated by the following formula:

$$\overline{D}_k = \frac{V_k}{bL_k} \tag{A.4.2-4}$$

where

$\overline{D}_k$ is the average deposition thickness of basin section k (m);

b is the working width of basin section k (m).

The deposition thickness of each cross section in the time frame in front of the overflow area may be calculated by cross sections from downstream to upstream, starting from the downstream section of basin section $k = n$ (cross section $n+1$).

3 The deposition volume of the whole basin chamber in a flushing cycle can be calculated by the following formulae:

$$V_s = \sum_{j=1}^{m_0} V_j \qquad (A.4.2\text{-}5)$$

$$V_j = \sum_{k=1}^{n} V_k + V_f \qquad (A.4.2\text{-}6)$$

$$V_f = b_f L_f D_{n+1} \qquad (A.4.2\text{-}7)$$

where

V_s is the accumulated deposition volume in each time frame of the whole basin chamber (m^3);

V_j is the deposition volume in a given time frame of the whole basin chamber (m^3);

V_f is the deposition volume in overflow area in a given time frame (m^3);

L_f is the length of overflow area (m);

D_{n+1} is the time-frame deposition thickness of the cross section $n+1$ (m);

j is the numbering of time frames, $j = 1, 2, \ldots, m_0$.

4 The basin chamber shall be flushed in time when the sediment concentration and particle size at the outlet exceed the design values, and the flushing cycle of the basin chamber can be calculated by the following formula:

$$T_0 = \sum_{j=1}^{m_0} t_j \qquad (A.4.2\text{-}8)$$

A.4.3 The settling rate of the desilting channel may be calculated in accordance with Section A.2.1 of this code. The time-frame deposition volume of each basin section can be calculated by Formula (A.4.2-1) of this code, and the section deposition thickness in the time frame may be calculated by sections from downstream to upstream, starting from the end section of the basin. The calculation shall meet the following requirements:

1 The deposition volume of the whole basin in the jth time frame can be calculated by the following formula:

$$V_j = \sum_{k=1}^{n} V_k \qquad (A.4.3\text{-}1)$$

2 The deposition volume of the whole basin from the 1th to m_0th time frame can be calculated by the following formula:

$$V_s = \sum_{j=1}^{m_0} V_j \tag{A.4.3-2}$$

Appendix B Calculation on Suspended Sediment Settling Velocity

B.0.1 Particle sizes should be classified by the Φ method. The classification shall be able to govern the particle size distribution curve, and shall include the minimum design particle size for sedimentation.

B.0.2 If the particle size distribution of suspended load is obtained by the method of sediment settling analysis, the sediment settling velocity shall be calculated by the formulae specified in the method.

B.0.3 For the suspended load, the particle size finer than 0.062 mm obtained by the method of sediment settling analysis, and the particle size larger than 0.062 mm obtained by the sieving method, the sediment settling velocity may be calculated as follows:

1 When the particle size is equal to or finer than 0.062 mm, the settling velocity can be calculated by the following formulae:

$$\omega = \frac{g}{1800}\left(\frac{\rho_s - \rho_w}{\rho_w}\right)\frac{d^2}{v} \qquad (B.0.3\text{-}1)$$

$$v = \frac{0.01775}{1 + 0.0337t + 0.000221t^2} \qquad (B.0.3\text{-}2)$$

where

- ω is the settling velocity of sediment (cm/s);
- g is the acceleration of gravity (cm/s^2);
- ρ_s is the sediment density (t/m^3);
- ρ_w is the clear water density (t/m^3);
- d is the particle size of sediment (mm);
- v is the kinematic viscosity coefficient of water (cm^2/s);
- t is the water temperature (°C).

2 When the particle size ranges from 0.062 mm to 2.000 mm, the settling velocity can be calculated by the following formulae:

$$\omega = S_a\left[gv\left(\frac{\rho_s}{\rho_w} - 1\right)\right]^{1/3} \qquad (B.0.3\text{-}3)$$

$$(\log S_a + 3.665)^2 + (\log \psi - 5.777)^2 = 39.0 \qquad (B.0.3\text{-}4)$$

$$\psi = \frac{\left[g\left(\dfrac{\rho_s}{\rho_w}-1\right)\right]^{1/3} d}{10\upsilon^{2/3}} \qquad (B.0.3\text{-}5)$$

where

S_a is the discriminant coefficient of settling velocity;

ψ is the discriminant coefficient of particle size.

3   When the particle size is larger than 2.000 mm, the settling velocity can be calculated by the following formula:

$$\omega = 0.557\left[gd\left(\dfrac{\rho_s}{\rho_w}-1\right)\right]^{1/2} \qquad (B.0.3\text{-}6)$$

B.0.4 When the design inflow sediment concentration is large, the settling velocity shall be corrected by an appropriate formula considering the effects of sediment concentration.

B.0.5 The average settling velocity of the particle size group shall take the geometric mean of the settling velocities of the upper and lower limit particle sizes, and be calculated by the following formula:

$$\overline{\omega}_i = \sqrt{\omega_i \omega_{i+1}} \qquad (B.0.5)$$

where

$\overline{\omega}_i$ is the average settling velocity of the particle size group (cm/s);

ω_i is the settling velocity of the lower limit of the particle size group (cm/s);

ω_{i+1} is the settling velocity of the upper limit of the particle size group (cm/s).

Appendix C Hydraulic Calculation on Sediment Releasing for Desilting Basin

C.1 Hydraulic Calculation of Pressure Sediment Releasing

C.1.1 The pressure sediment releasing channel shall not be silted up, the selected sediment-flushing flow velocity v_C shall be greater than the critical sediment-flushing flow velocity v_k and should not be less than 2.5 m/s. Critical sediment-flushing flow velocities of sediment releasing channels with rectangular and circular sections may be determined as follows:

1 For rectangular section:

$$v_k = E[(\rho_m - 1)\omega_{75}]^{1/3} R^{1/2} \tag{C.1.1-1}$$

$$\rho_m = 1 + \left(1 - \frac{\rho_w}{\rho_s}\right)\frac{S_e}{1000} \tag{C.1.1-2}$$

$$S_e = \eta S_0 \frac{Q + Q_c}{Q_c} \tag{C.1.1-3}$$

where

- v_k is the critical sediment-flushing flow velocity (m/s);
- E is the constant related to the absolute roughness Δ of the gallery surface, which is taken as 66 when Δ is 1 mm, and 50 when Δ is 5 mm;
- ρ_m is the density of muddy water in the gallery (t/m³);
- ω_{75} is the sediment settling velocity (m/s), where the weight of finer sediment accounts for 75 % in the sediment-laden flow of the gallery;
- S_e is the sediment concentration in flushing flow (kg/m³);
- S_0 is the design inflow sediment concentration (kg/m³);
- Q_c is the sediment flushing flow (m³/s).

2 For circular section:

$$v_k^{5/4} = \omega_{75}\left(\frac{100 S_e}{1000 + 0.63 S_e}\right)^{1/6}\left(\frac{4 Q_c}{\pi d_{75}^2}\right)^{1/4} \tag{C.1.1-4}$$

where

- d_{75} is the sediment particle size (m), where the weight of finer sediment accounts for 75 % in the sediment-laden flow of the gallery.

C.1.2 The head loss at the confluence of each branch gallery and the main gallery shall be equal, and the total head loss of the gallery shall be less than the water level difference between the upstream water surface and the outlet of the gallery. The flow velocity in each branch gallery should be the same, and the equal head loss may be realized by varying the length of each branch gallery. The water head change along the gallery can be calculated by the following formulae:

$$H_1 = \Delta Z - (1 + \Sigma \xi) \frac{v_k^2}{2g} - v_k^2 \int_0^{L_1} \frac{dl}{C^2 R} \qquad \text{(C.1.2-1)}$$

$$C = \frac{1}{n} R^{1/6} \qquad \text{(C.1.2-2)}$$

where

- H_1 is the water head of the calculation section (m);
- ΔZ is the water level difference between the upstream water surface and the gallery outlet (m);
- $\Sigma \xi$ is the sum of local head loss coefficients;
- L_1 is the length from the head of the gallery to the calculation section (m);
- g is the acceleration of gravity (m/s^2);
- n is the roughness coefficient ranging from 0.0170 to 0.0225, which is taken as 0.0225 when calculating the water head difference of the branch gallery, and 0.0170 when calculating the area of the feed hole.

C.1.3 The cross-sectional area of the branch gallery along the length can be calculated by the following formulae:

$$A = \frac{Q_1}{v_k} \qquad \text{(C.1.3-1)}$$

$$Q_1 = Q_0 + \frac{Q_m - Q_0}{L_T} L_c \qquad \text{(C.1.3-2)}$$

where

- A is the sectional area of branch gallery (m^2);
- Q_1 is the flow of the branch gallery at the calculation section (m^3/s);

Q_0　is the flow at the head of the branch gallery, which may be taken as 20 % of the total flow of the main gallery (m³/s);

Q_m　is the flow at the end of the branch gallery (m³/s);

L_T　is the total length of branch galleries (m);

L_c　is the length from the head of the branch gallery to the calculation section (m).

C.1.4 The cross-sectional area of the sediment feed hole of the branch gallery can be calculated by the following formula:

$$A_1 = \frac{Q_1}{n_0 \Phi \sin\theta \sqrt{2g(H-H_1)}} \tag{C.1.4}$$

where

A_1　is the cross-sectional area of sediment feed hole (m²);

n_0　is the number of sediment feed holes;

θ　is the inclination angle between the hole axis and the horizontal plane (°);

Φ　is the flow coefficient ranging from 0.75 to 0.85, which may be taken as 0.75 for louver holes.

C.2 Hydraulic Calculation of Non-pressure Sediment Releasing

C.2.1 The critical flow velocity of non-pressure sediment releasing and flushing can be calculated by the following formula:

$$v_k = \omega_{75}\sqrt{\frac{\overline{h}}{d_{75}}}\left(\frac{100S_e}{1000+0.63S_e}\right)^{1/4} \tag{C.2.1}$$

where

$\overline{h}$　is the average flow depth during sediment flushing (m);

S_e　is the sediment concentration in the flushing flow, which may be taken as 20 kg/m³ to 85 kg/m³.

C.2.2 The sediment flushing flow per unit width can be calculated as follows:

$$q = (1.1 \sim 1.25) h v_k \tag{C.2.2}$$

C.2.3 The duration of sediment flushing in constant velocity can be calculated by the following formula:

$$\Delta T = (1.5 \sim 2.0)\frac{1000 \rho_d V}{S_e q B} \tag{C.2.3}$$

where

ΔT is the sediment flushing duration (s);

V is the deposition volume of the desilting basin (m³);

B is the working width (m).

Appendix D Hydraulic Calculation on Sediment Flushing for Desilting Basin

D.0.1 For multi-chamber desilting basin, each basin chamber may settle and flush alternately. Under the design condition, one of the basin chambers should be used for flushing while others for settling, so the basin chambers operate alternately to maintain continuous water supply. The flushing process (Figure D.0.1) may be divided into three stages: evacuating flushing, backward flushing and flushing along the course.

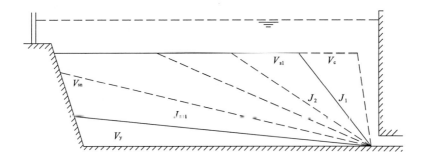

Key

V_c — is the sediment releasing volume for evacuating flushing

$V_{s1}, ..., V_{sn}$ — is the sediment releasing volume for backward flushing

V_y — is the sediment releasing volume for flushing along the course

$J_1, ..., J_{n+1}$ — is the starting and ending slopes of ith calculation flushing block

Figure D.0.1 Flushing process

D.0.2 The sediment releasing volume and duration of basin chamber evacuating flushing may be determined as follows:

1 For sediment releasing volume:

$$V_c = \frac{V_w S_c}{1000\rho_d - S_c} \tag{D.0.2-1}$$

where

V_c is the sediment releasing volume (m³);

V_w is the water volume above the initial siltation surface (m³);

S_c is the average outflow sediment concentration (kg/m³).

2 For flushing duration:

$$t_c = \frac{V_w + V_c}{3600 Q_c} \tag{D.0.2-2}$$

where

  t_c is the flushing duration (h);

  Q_c is the average evacuating flow (m³/s).

D.0.3 The calculation of backward flushing of basin chamber should take the basin bottom in front of the flushing sluice as the starting point, and generalize the sediment volume into several calculation flushing blocks, and each block may be calculated as follows:

1 The average sediment transport rate per unit width of each flushing block can be approximately calculated by the following formulae:

$$q_{si} = \beta_s q_i^{m_s} \overline{J}_i^{n_s} \tag{D.0.3-1}$$

$$\overline{J}_i \leq \frac{J_i + J_{i+1}}{2} \tag{D.0.3-2}$$

where

  q_{si} is the average sediment transport rate per unit width of the calculation flushing block [kg/(s · m)];

  q_i is the average flushing flow per unit width of the calculation flushing block [m³/(s · m)];

  $\overline{J}_i$ is the average slope of the block (%);

  J_i, J_{i+1} are the starting and ending slopes of the block (%);

  i is the numbering of the block, $i = 1, 2, ..., n$;

  β_s is the comprehensive coefficient of the backward flushing;

  m_s, n_s are the indexes of backward flushing.

2 For backward flushing duration:

$$t_s = \frac{1}{3.6} \sum_{i=1}^{n} \left(\frac{V_{si} \rho_d}{q_{si} b_i - S_0 Q_s} \right) \tag{D.0.3-3}$$

where

  t_s is the backward flushing duration (h);

  V_{si} is the sediment volume of the block (m³);

  b_i is the average width of the calculation block (m);

  S_0 is the inflow sediment concentration (kg/m³);

  Q_s is the flushing flow (m³/s).

D.0.4 The average outflow sediment concentration and the duration of the flushing along the course may be determined as follows:

1 For average outflow sediment concentration:

   1) When the measured data of existing desilting basins with similar flow and sediment characteristics are available, the average outflow sediment concentration of the flushing along the course can be calculated by the following formula:

$$S_* = K_* \left(\frac{v_*^3}{gR_*\omega_{10}} \right)^{m_*} \qquad (D.0.4\text{-}1)$$

where

S_* is the average outflow sediment concentration (kg/m³);

v_* is the average flow velocity (m/s);

R_* is the average hydraulic radius (m);

g is the acceleration of gravity (m/s²);

ω_{10} is the average settling velocity of sediment particle size larger than d_{10} (m/s);

K_* is the coefficient;

m_* is the index.

   2) In the absence of measured data of existing desilting basins with similar flow and sediment characteristics, the average outflow sediment concentration of the flushing along the course can be calculated by the following formulae:

$$S_* = 2.5 \left[\frac{(0.0022 + S_V)U^3}{\kappa_m \frac{\rho_s - \rho_m}{\rho_m} gh\omega_s} \ln\left(\frac{h}{6D_{50}}\right) \right]^{0.62} \qquad (D.0.4\text{-}2)$$

$$S_V = \frac{S_0}{2700} \qquad (D.0.4\text{-}3)$$

$$\omega_s = \omega_{cp}(1 - 1.25S_V)\left(1 - \frac{S_V}{2.25\sqrt{d_{50}}}\right)^{3.5} \qquad (D.0.4\text{-}4)$$

$$\kappa_m = \kappa_0 \left[1 - 4.2\sqrt{S_V}(0.365 - S_V)\right] \qquad (D.0.4\text{-}5)$$

$$\rho_{\mathrm{m}} = \rho_{\mathrm{w}} + \left(\frac{\rho_{\mathrm{s}} - \rho_{\mathrm{w}}}{\rho_{\mathrm{s}}}\right)\frac{S_0}{1000} \tag{D.0.4-6}$$

where

S_V is the specific volume sediment concentration (%);

ω_{s} is the sediment settling velocity in muddy water (m/s);

ω_{cp} is the average sediment settling velocity in clear water (m/s);

ρ_{m} is the density of muddy water (t/m³);

h is the flow depth (m);

D_{50} is the median particle size of bed sediment (m), which may be approximately taken as 2 to 3 times d_{50} in the absence of measured data;

d_{50} is the median particle size of suspended load (mm);

κ_{m} is the Carmen constant of muddy water;

κ_0 is the Carmen constant of clear water, which should be taken as 0.4.

2   For duration of flushing along the course:

$$t_{\mathrm{y}} = \frac{\rho_{\mathrm{d}} V_{\mathrm{y}}}{3.6 Q_{\mathrm{s}}(S_* - S_0)} \tag{D.0.4-7}$$

where

t_{y} is the flushing duration (h);

V_{y} is the sediment releasing volume (m³).

D.0.5 The total duration of single chamber flushing can be calculated by the following formulae:

$$T_{\mathrm{c}} = K(t_{\mathrm{c}} + t_{\mathrm{s}} + t_{\mathrm{y}}) \tag{D.0.5-1}$$

$$T_{\mathrm{c}} \leq \frac{24 T_0}{n_{\mathrm{c}} - 1} \tag{D.0.5-2}$$

where

T_{c} is the total duration of single chamber flushing (h);

K is the working safety factor, which may be taken as 1.2 to 1.4;

n_{c} is the number of basin chambers.

D.0.6 The total water consumption for single chamber flushing can be calculated by the following formula:

$$W = V_w + 3600(t_s + t_y)Q_s \tag{D.0.6}$$

where

W　is the total water consumption for single chamber flushing (m³).

D.0.7 When basin tanks are set in a basin chamber, the evacuating flushing should be calculated as one-time evacuating flushing of the basin chamber. Backward flushing and flushing along the course may be calculated tank by tank using the formula for basin chamber calculation.

Explanation of Wording in This Code

1   Words used for different degrees of strictness are explained as follows in order to mark the differences in executing the requirements in this code.

    1) Words denoting a very strict or mandatory requirement:

    "Must" is used for affirmation, "must not" for negation.

    2) Words denoting a strict requirement under normal conditions:

    "Shall" is used for affirmation, "shall not" for negation.

    3) Words denoting a permission of a slight choice or an indication of the most suitable choice when conditions permit:

    "Should" is used for affirmation, "should not" for negation.

    4) "May" is used to express the option available, sometimes with the conditional permit.

2   "Shall meet the requirements of…" or "shall comply with…" is used in this code to indicate that it is necessary to comply with the requirements stipulated in other relative standards and codes.

List of Quoted Standards

GB/T 50159,	*Code for Measurement of Suspended Load in Open Channels*
NB/T 35023,	*Design Code for Sluice*
DL/T 5057,	*Design Specification for Hydraulic Concrete Structures*
SL 616,	*Specification for Hydraulic Prototype Observation*